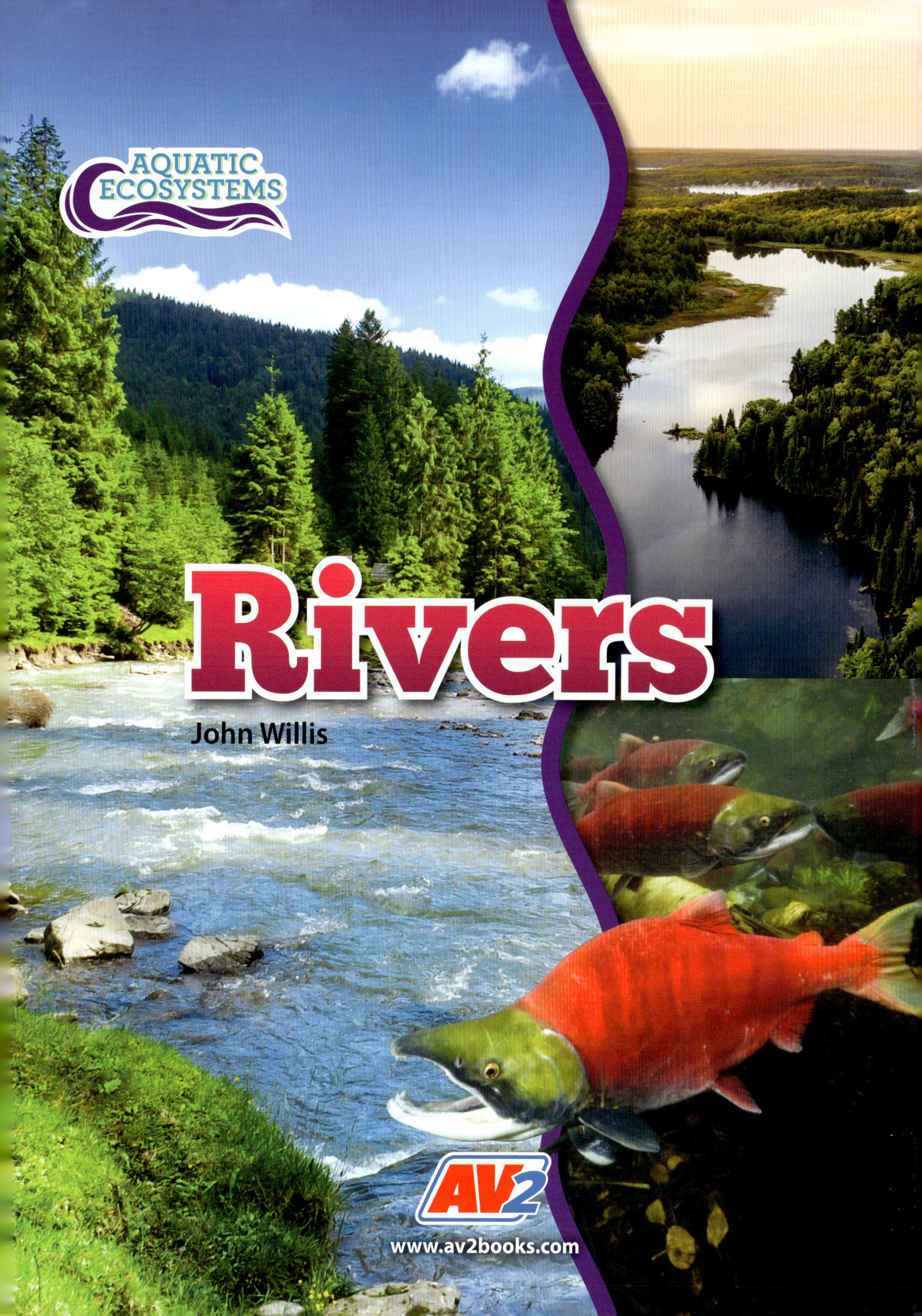
AQUATIC ECOSYSTEMS
Rivers
John Willis
AV2
www.av2books.com

Step 1
Go to **www.av2books.com**

Step 2
Enter this unique code
AFEJOOGNK

Step 3
Explore your interactive eBook!

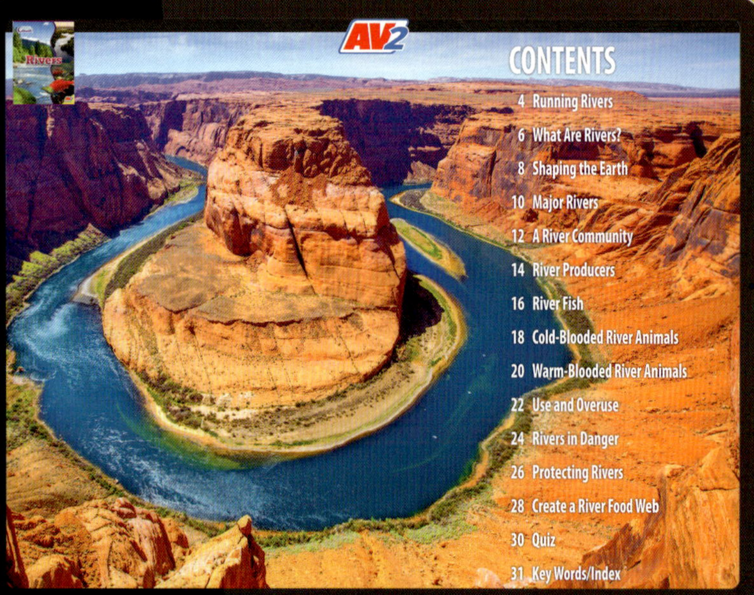

AV2 is optimized for use on any device

Your interactive eBook comes with...

Contents
Browse a live contents page to easily navigate through resources

Audio
Listen to sections of the book read aloud

Videos
Watch informative video clips

Weblinks
Gain additional information for research

Try This!
Complete activities and hands-on experiments

Key Words
Study vocabulary, and complete a matching word activity

Quizzes
Test your knowledge

Slideshows
View images and captions

... and much, much more!

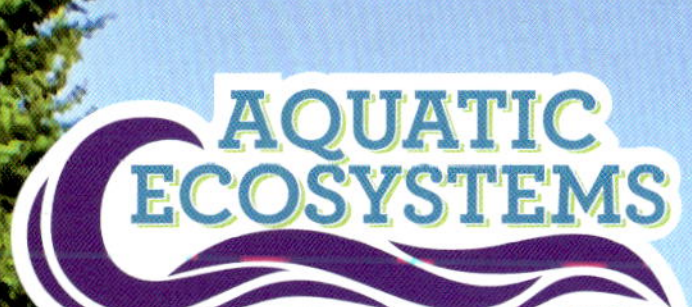

Rivers

Contents

Running Rivers

Rivers have been important to humans since before recorded history. They provide food, water, and healthy soil for crops. Many of the earliest known human **settlements** were found on riverbanks. Over time, civilizations grew around these bodies of water. Today, some of the world's largest cities are found near rivers.

Rivers do not include a large amount of Earth's water. However, they are an important source of fresh water for both humans and animals. Many different fish and other animals either live in rivers or depend on them for food and water. This makes rivers a key aquatic **ecosystem**. Rivers also play a major role in returning water from land to the oceans.

Record-Setting Rivers

Highest

The **Yarlung Tsangpo River** is **14,800 feet** (4,500 meters) above sea level at its highest point.

Deepest

The **Congo River** is **720 feet** (219 m) deep at its deepest point.

Largest Flow Volume

7,380,765 cubic feet (209,000 cubic meters) of water flows out of the **Amazon River** every second.

Only about 0.025 percent of Earth's water is found flowing in rivers at any time.

What Are Rivers?

Rivers are bodies of moving water. They travel downhill in a **depression** known as a channel. Rivers are known by many other names, including streams and creeks. Often, these words are used to describe smaller kinds of rivers. The water in rivers comes from many different sources, such as glaciers, precipitation, and springs. Smaller rivers that flow into larger ones are known as tributaries. Although they may have different names, all rivers are made of moving water.

Rivers fed by glaciers often have little or no flowing water during colder seasons.

Rivers are made of several main parts. All rivers have a starting point where water begins its flow. This source is called a headwater. From the headwater, a river flows along its channel. There are many different kinds of channels. Their shape and depth depend on the type of soil or rock that the river flows through. Some channels are wide and straight. Others bend or split.

The land along the side of a channel is known as a riverbank. A river's end, where it reaches another body of water, is known as its mouth. This is usually the widest part of the river. Often, small particles of soil are carried by rivers and build up at the mouth. They form a fan-shaped area of land known as a delta.

Most deltas form where rivers reach lakes or oceans. Inland deltas, where a river empties into the land, are much rarer.

Longest Rivers by Continent

Africa
Nile River
4,160 miles
(6,695 kilometers) long

South America
Amazon River
3,976 miles
(6,400 km) long

Asia
Yangtze River
3,915 miles
(6,300 km) long

North America
Mississippi River
2,340 miles
(3,766 km) long

Europe
Volga River
2,193 miles
(3,530 km) long

Australia
Murray River
1,570 miles
(2,350 m) long

Antarctica
Onyx River
20 miles
(32 km) long

Shaping the Earth

Rivers change the world around them, altering and creating new habitats for people and animals. When water in a river moves, it **erodes** the land around it. Sometimes, this causes a river's channel to move or change shape. Erosion can also make riverbanks steeper as small amounts of material are carried downstream. Over long periods of time, this can cause land features such as canyons to form. The Grand Canyon, found in the Southwest United States, is more than 6,000 feet (1,829 m) deep in certain areas. It was created by the Colorado River over the last 6 million years.

Rivers can also cause Earth's surface to change quickly. Heavy rains or the melting of glaciers in spring can cause rivers to swell and rise above their banks. This can cause flooding downstream. Some floods happen seasonally. Often, these regular floods are helpful. They enrich the soil in the areas around the river. However, when floods are caused by unexpected events, such as many rainstorms in a short period of time, they can become natural disasters. In 1927, the Mississippi river flooded, covering more than 23,000 square miles (59,570 square km) of land. Hundreds of people died and hundreds of thousands more lost their homes.

Rivers travel along the path of least resistance. Over time, this can cause large bends to form, such as the Grand Canyon's Horseshoe Bend.

Floods that occur very suddenly are known as flash floods. They are responsible for more deaths than other kinds of floods.

Major Rivers

Rivers are found on every continent. They include many kinds of **environments**. Rivers are homes for many animals. They are also key transportation routes for humans. Some rivers are known for their historical importance. Others have been threatened by human activity in recent years.

Mississippi River

The Mississippi River flows entirely within the United States. It has been an important transportation route for much of the nation's history. Today, it is still one of the busiest rivers in the world. More than 175 million tons (159 million metric tons) of freight pass through the Mississippi River every year.

North America
Atlantic Ocean
EQUATOR
Pacific Ocean
South America

Legend
Land
Major River
Water
N
S
E
W
Scale
0
2,000 Miles
2,000 Kilometers

Nile River

The Nile flows through or beside 11 countries in Africa, including Egypt. Egyptians have relied on this river for water and rich soil for thousands of years. Today, 95 percent of people in Egypt still live close to the Nile.

Ganges River

More than 400 million people depend on the Ganges River for water. However, in recent years, this river ecosystem has become very polluted. Sewage and **runoff** from towns and farms have filled it with bacteria and dangerous chemicals. Today, the Indian government has put programs in place to help clean the Ganges.

Asia

Europe

Pacific Ocean

Indian Ocean

Australia

Southern Ocean

Amazon River

The area of land that drains into the Amazon River is known as the Amazon River Basin. It covers about 38 percent of South America. This basin is one of the most **diverse** ecosystems on Earth. It is home to millions of different species of insects, birds, and other animals. The river itself has more than 5,600 species of fish.

A River Community

There are different river ecosystems around the world. However, they all share some features. In every ecosystem, different **organisms** occupy certain roles. Energy passes between these plants and animals when they eat or are eaten.

Organisms called producers receive energy from the Sun. They turn this energy into a form that they can use. When other organisms eat producers, they take this energy for themselves. When an organism that eats a producer is eaten, the energy is passed on again. This means that all organisms in an ecosystem need producers for their energy.

Primary consumers are animals that eat producers. Most primary consumers are **herbivores**. They are eaten by both secondary and tertiary consumers. These are usually **carnivores** or **omnivores**. Often, a tertiary consumer has no **predators**. When organisms die, **decomposers** break down their remains. This recycles their energy. It turns into nutrients that producers can use.

Every year, thousands of bald eagles gather at Alaska's Chilkat River to feed on salmon.

River Energy Pyramid

As energy passes through an ecosystem, some of it is lost. This is why there are fewer consumers than producers in an ecosystem. This relationship can be shown through an energy pyramid.

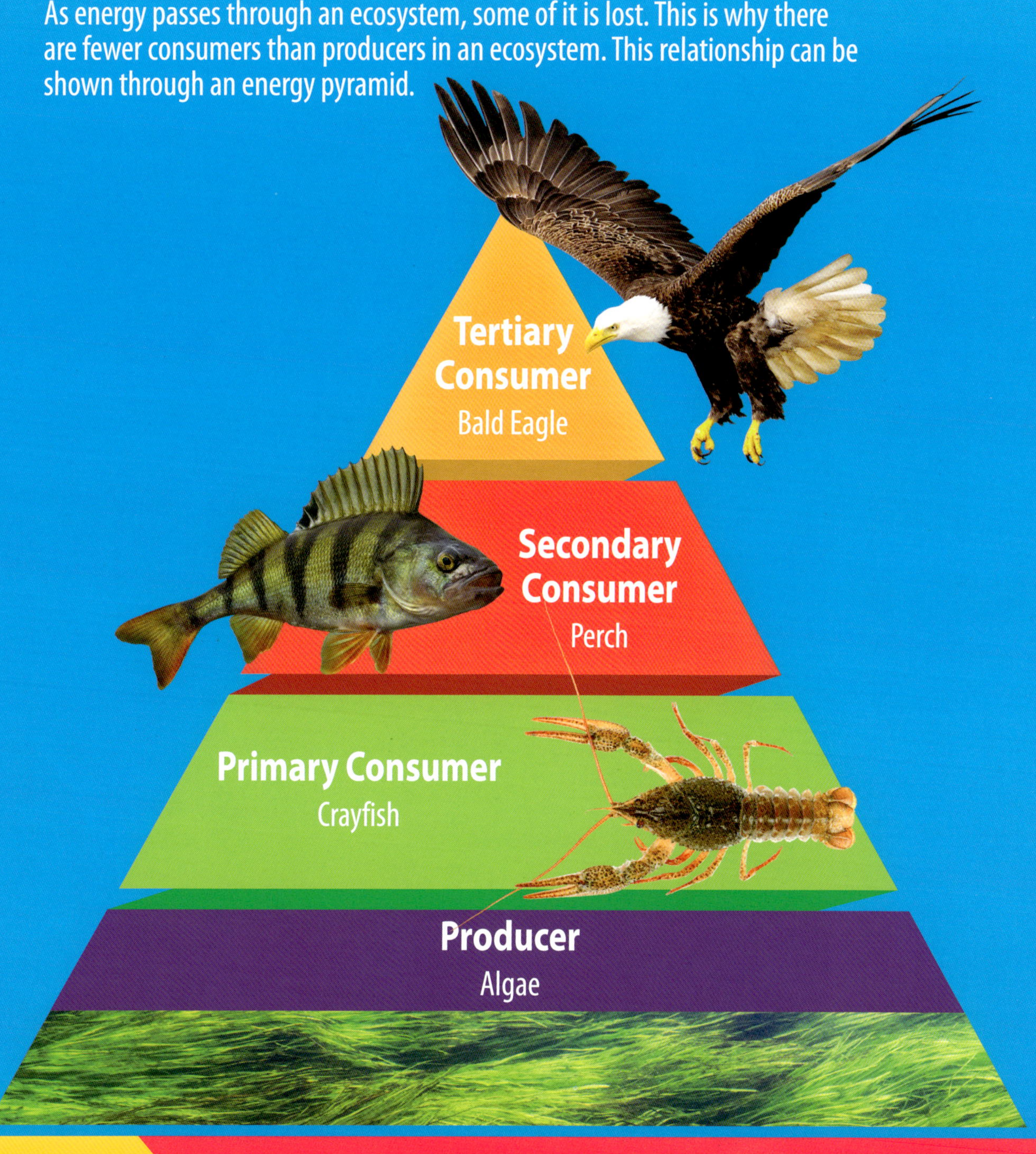

Try This! Choose a river plant or animal. Research it using the library or the internet. Is it a producer or a consumer? If it is a consumer, what kind is it? Use your research to create an energy pyramid that includes your chosen organism.

River Producers

Most river producers are plants or **algae**. River plants, like all producers, are an essential part of a river ecosystem. They turn the Sun's energy into a form that all other living things in the ecosystem depend on. Plants that live in water have unique **adaptations** that allow them to survive. Underwater plants are often more flexible than land plants. They use the water itself to keep their shape. Plants that live in deeper waters also need to survive with less sunlight than those on the land.

Water plants can greatly affect the appearance of a river. Caño Cristales, in Colombia, is known as the "river of five colors" due in part to the red plants living in it.

Living in a river can present even more problems to plants. River plants need to be able to survive in moving water. A fast-flowing river could tear or uproot a plant. Some river plants have very strong roots. These anchor them to the riverbed and stop them from getting washed downstream. River plants may have specially adapted leaves as well. A reed's long, thin leaves move with the water. This makes them less likely to tear. Algae are the most common kind of river producer. In slow rivers, algae float along the surface. In faster rivers, algae grow on rocks.

River plants are able to get large amounts of nutrients from the water around them. This is because nutrients upstream will move downstream with the water. However, this also makes river plants vulnerable to pollution. This can include waste from sewer systems and runoff from farms. These pollutants can affect river plants far downstream from their origins.

Algae are not true plants. They do not have leaves, stems, or roots. However, like plants, they produce energy using sunlight.

River Fish

Rivers are home to many different kinds of fish. The moving water in a river or stream can be helpful to them. Fish use organs called gills to take oxygen from the water. In clean, running water, a fish can absorb up to 85 percent of the oxygen in the water that flows past its gills.

Many fish that live in rivers have narrow, streamlined bodies. This shape cuts through the water, allowing them to swim upstream more easily. Other river fish stay out of strong currents by living on the bottom of a river. These fish have bodies that float less easily than fish that live closer to the top of the water.

Catfish spend most of their lives near the bottom of a body of water. Many types, such as channel catfish, can be found in deep rivers. Catfish have long whiskers called barbels. A catfish's body and barbels are covered in taste buds. This allows the fish to feel for prey in the dark waters at a river's bottom. Catfish are secondary consumers. They eat smaller fish and other animals. Larger fish and some birds eat them.

Catfish were given their name because the barbels on their faces look like cat whiskers.

Some fish spend only part of their lives in rivers. Salmon are born in rivers. Young salmon have dark gray stripes on their bodies. This helps them blend in with the rocks in river ecosystems. Some young salmon spend about two years in fresh water, while others immediately head downstream to the ocean. Once salmon reach the ocean, their bodies change. They turn silver to blend in with their new home. Their organs change to help them survive in **salt water**.

When adult salmon are ready to lay eggs, their bodies change again to allow them to live in fresh water. Most do not eat, using all of their energy to swim upstream. During this time,many animals, such as birds and bears, prey upon them. Once salmon reach the place where they were born, they lay eggs.

Some salmon turn red once they have returned to the place where they were born. This color shows that they are ready to reproduce.

Cold-Blooded River Animals

A cold-blooded animal is one that cannot maintain a constant body temperature on its own. Instead, it must rely on its surroundings to cool off or warm itself. These animals are also known as ectotherms. Many fish, along with amphibians, reptiles, and insects, are ectotherms. These animals can often be found in river ecosystems.

River Amphibians

Amphibians are able to breathe through their smooth skin. All amphibians need water in order to avoid drying out, so most are found living in or near it. Amphibians that live in rivers are uniquely adapted to their ecosystem. Because amphibians breathe through their skin, they are most often found in clean rivers.

Tailed frogs live in fast-flowing rivers. Their young use suckers to attach to rocks in moving water. Adults use sharp claws for the same purpose. Tailed frogs have smaller lungs than most other frogs. This helps them stay near the bottom of a river, away from strong currents.

A tailed frog's body color typically matches the rocks in its home.

River Reptiles

Unlike amphibians, reptiles do not need to keep their skin wet. However, many of these scaled animals, including crocodiles and turtles, still live in rivers. In order to keep warm, reptiles leave the water to bask in the Sun.

Nile crocodiles grow to be about 16 feet (4.9 m) long.

Turtles that live in fast-flowing rivers, such as Asia's big-headed turtle, use their strong claws and beaks to grip rocks. This allows them to stay in one place underwater without having to swim against the current. Nile crocodiles have colors that allow them to hide in rivers. Their eyes and nostrils are on the top of their head, allowing them to breathe and see while staying mostly underwater. Nile Crocodiles are tertiary consumers. They **ambush** fish and animals that cross or drink from a river.

River Insects

Insects are six-legged animals found throughout the world. Most insects are small and light. They need special adaptations to live in the flowing water of a river ecosystem.

Insects may only spend part of their lives in rivers. Young caddisflies live in fast-flowing rivers. Many use pieces of rock or wood to build a protective casing. Others make webs out of silk. These webs catch food as it flows downstream. Caddisflies leave the water when they become adults. Young mayflies live in streams, using gills to breathe. They spend most of their lives underwater. When they become adults, they leave the water and fly to find a mate. Some mayflies only live as adults for a few hours.

Adult caddisflies are known for their moth-like appearance.

Warm-Blooded River Animals

A bear may eat more than 30 salmon in a single day.

Warm-blooded animals can maintain a constant body temperature. They are also known as endotherms. Endotherms do not depend on their environment to keep warm or cool. Both birds and mammals are endotherms. Many of these animals live near rivers or rely on them for food.

River Mammals

Some mammals in a river ecosystem spend most of their time in the water. River otters have torpedo-shaped bodies that help them cut through water. Their strong tails and webbed feet propel them when they swim.

Other mammals in river ecosystems live on land but rely on rivers for food. Every summer, many bears gather in rivers in North America. As salmon swim upstream to lay eggs, the bears hunt. These bears rely on the fat from salmon in order to **hibernate** through the winter. In areas where rivers have been blocked, bears may not be able to find the salmon they need.

Beavers have a unique way of living in rivers. These large rodents are able to change their ecosystem completely. They block rivers by cutting down trees to make dams. This can both help and harm other animals. Beaver dams may change the flow of a river ecosystem or turn it into a different body of water, such as a pond or wetland. This creates new homes for animals that rely on still waters to survive.

River Birds

Most river birds do not spend their lives living in the water. However, rivers are an important source of food for many birds. Large birds are often tertiary consumers in river ecosystems, feeding on fish and other animals.

Kingfishers live near bodies of water, including rivers. They wait on branches above the water, then dive and use their sharp beaks to catch fish. Kingfishers can see through reflections on the water's surface. This allows them to spot fish more easily.

The American dipper is the only songbird in North America that hunts underwater. These robin-sized birds swim or walk along the bottom of rivers, looking for food under rocks. Their thick coat of waterproof feathers help them survive in cold rivers.

There are about 90 different species of kingfishers.

Use and Overuse

The Euphrates River is 1,740 miles (2,800 km) in length.

Rivers are important ecosystems for animals. However, humans also rely on them in many different ways. The Fertile Crescent is an area in the Middle East that is also nicknamed the "Cradle of Civilization." It includes parts of the Nile, Tigris, and Euphrates rivers. The soil in this area has historically been good for growing crops. The Fertile Crescent gets its nickname because scientists believe humanity first began to move from hunting and gathering to farming there. In addition to agriculture, the Fertile Crescent is also the site at which science, writing, and many other parts of human culture began. Scientists believe this is because the river allowed people to stay in one place.

In more recent times, people have used rivers in many new ways. They can also be a source of power. Rivers powered mills used to grind grains. More recently, hydroelectric dams have allowed people to use flowing water found in rivers to power buildings. These dams hold back river water in a reservoir. When water is allowed to flow through the dam, it passes by a turbine. As the turbine spins, it generates electricity.

People often use rivers for leisure as well. Fishing is a common leisure activity throughout the world. Some rivers, such as Europe's Rhône and Rhine, are large enough for cruise ships. In smaller rivers, people use canoes or kayaks to travel downstream. Stronger-flowing rivers are enjoyed by whitewater rafters. They enjoy the challenge of navigating rapids and currents.

River Rapid Categories

Whitewater rafters use the International Scale of River Difficulty in order to determine if they can safely travel through a river's rapids.

Class I Easy	Fast-moving water with obvious obstacles. Easy for swimmers to self-rescue.
Class II Moderate	Wide rapids with easy-to-avoid rocks and waves.
Class III Moderately difficult	Moderate, irregular waves. Often requires complex maneuvers.
Class IV Difficult	Powerful but predictable rapids. Requires complex maneuvers and tight boat handling.
Class V Extremely difficult	Very long, violent rapids. Often have complex, difficult routes.
Class VI Extraordinarily difficult	Unpredictable and very dangerous. Should only be rafted by experts under the best circumstances.

Think About It! People may want to change rivers in order to make them safer or more exciting for leisure activities. What effect could this have on a river ecosystem? Who or what would benefit? Who or what could be harmed?

Rivers in Danger

River ecosystems give people easy access to crops and transportation. However, chemicals and sewage from cities and farms can heavily pollute rivers. By the 1960s, many rivers around the world were very polluted. In 1969, Ohio's Cuyahoga River caught fire. The fire was caused by flammable oil and other pollutants. While the river had caught fire before, the 1969 fire caught the nation's attention. New laws were passed to keep the river clean. By 2019, fish from the river were considered safe to eat.

In more recent years, other rivers around the world have become polluted as nearby cities have grown larger. In Indonesia, the Citarum River is very important to the nation's economy. More than 2,800 factories rely on the river. However, many factories return unclean water back into the river. Today, the Citarum is known as one of the most polluted rivers on Earth. The government has now enacted a plan to educate the public about the dangers of throwing garbage in the river.

Rivers can be threatened in other ways beyond pollution. The Rio Grande River is in the United States and Mexico. Along the river's course, much of the water is taken for irrigation. Farms and cities such as Albuquerque need this water to grow crops. However, so much water has been taken that, in 2001, the river failed to reach the Gulf of Mexico for the first time. This has greatly affected animals in the ecosystem. For example, the Rio Grande silvery minnow was once found throughout the river. Today, it is found in less than 10 percent of its original range.

Water from the Rio Grande is often diverted using structures such as ditches and canals.

The Indonesian government's goal is to have water from the Citarum River be considered drinkable again by 2025.

Protecting Rivers

As river ecosystems are important to both humans and the environment, people have taken many steps to protect them. The Environmental Protection Agency (EPA) was founded after the 1969 Cuyahoga River fires. Laws such as the Clean Water Act also came into effect. These made it easier to stop pollutants from being dumped in U.S. rivers.

Mississippi River Timeline

1541

Spanish explorer Hernando de Soto becomes the first European to see the Mississippi River. European settlers become interested in the rich soil and other resources found in the area.

Early 1800s

Many steamboats are used to move both passengers and cargo along the Mississippi River. These boats are able to travel at speeds of 5 miles (8 km) per hour.

1924

The Upper Mississippi River Wildlife and Fish Refuge is established. Covering parts of four states, the refuge includes 240,000 acres (97,124 hectares) of important habitats for fish, plants, and birds in the Mississippi River's floodplains.

The balance between using a river and preserving its ecosystem can be very difficult. One example has been the Mississippi River, one of the most important rivers in the United States. During the nation's history, the river has changed in many ways. Some of these changes have harmed the plants and animals living in and around it. Others have helped to protect and preserve its ecosystem. To this day, many different groups try to balance the need to use this ecosystem with the need to protect it.

How do you think the Mississippi River will change in the future? What will cause these changes? Will these changes be positive or negative?

1938–1942

The Upper Mississippi River Wildlife and Fish Refuge is greatly affected by the construction of new dams designed to allow ships to travel up and down the river more easily. Many plants, along with the animals that depended on them, died or left.

1986

Congress recognizes the Mississippi River as an important ecosystem and transportation route. It establishes the Environmental Management Program (EMP) to balance protection and commerce.

2019

The Mississippi River experiences its longest-lasting flood in 90 years. Record-breaking rains cause more than 250,000 acres (101,171 hectares) of farmland to be ruined.

Create a River Food Web

Plants and animals in a river ecosystem rely on each other for food. This can be modeled using a food web. You can create a food web using the steps below.

1. Find at least three river organisms of each type—producers, primary consumers, secondary consumers, and tertiary consumers.

2. Write down the name of one producer. Then, add the name of a primary consumer that eats it. Use an arrow to show the transfer of energy.

3. Write down the name of a secondary consumer that eats your primary consumer and the name of a tertiary consumer that eats your secondary consumer. Use arrows to show the flow of energy. This will create a food chain.

4. One at a time, add each of the remaining organisms to your food chain. Add new arrows to show which organisms they eat or are eaten by. Once all the organisms have been added, your food chain will be a food web.

Starter Food Web

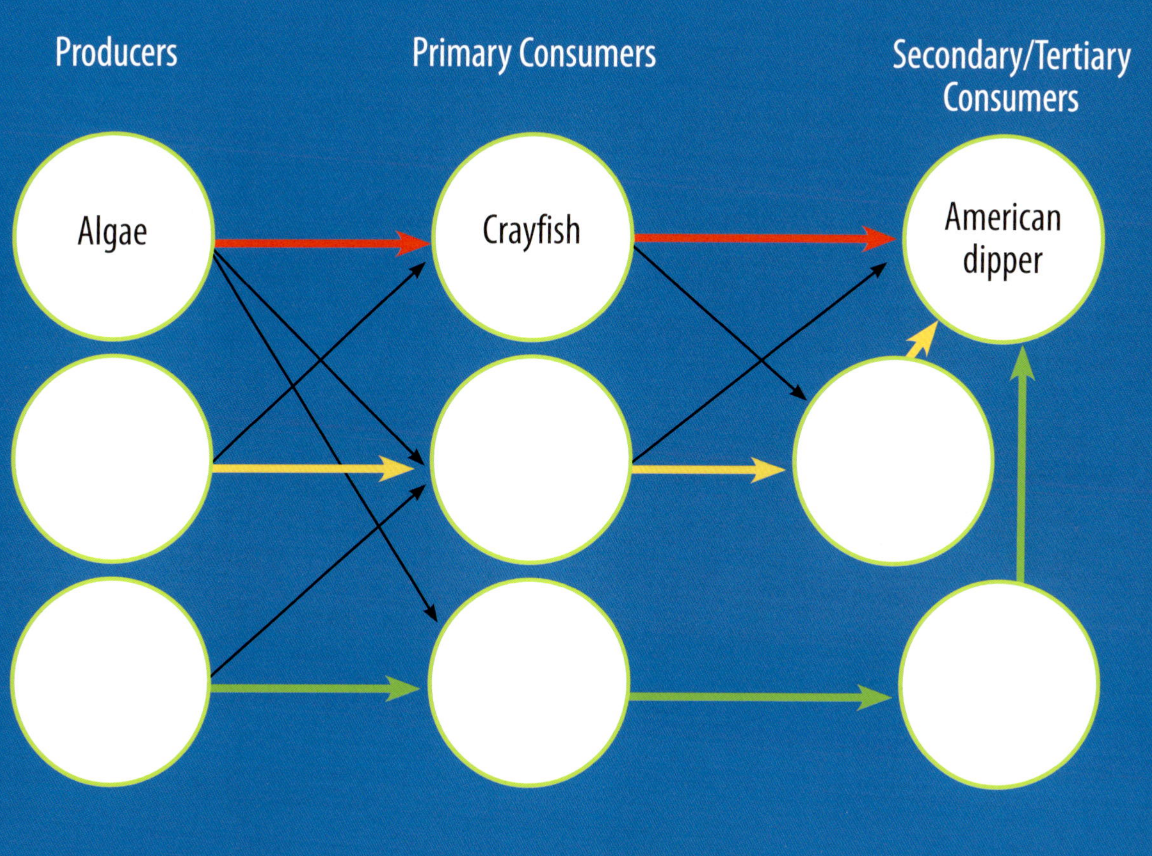

Once your food web is complete, use it to answer the following questions.

1. How would removing one organism from your food web affect the other organisms in the web?
2. What would happen to the rest of the food web if the producers were taken away?

Quiz

1 What are the whiskers on a catfish called?

2 How many countries does the Nile River flow through or beside?

3 Which river has the largest flow volume?

4 Which river in Indonesia is one of the most polluted rivers on Earth?

5 Is a bald eagle a secondary consumer?

6 What area includes the Nile, Tigris, and Euphrates rivers?

7 Is algae a kind of plant?

8 How many species of kingfisher are there?

ANSWERS

1 Barbels **2** 11 **3** The Amazon River **4** The Citarum River **5** No **6** The Fertile Crescent **7** No
8 About 90

Key Words

adaptations: changes that help an organism survive in its environment

algae: plantlike organisms that produce energy from sunlight

ambush: a surprise attack

carnivores: animals that eat only meat

decomposers: organisms that break down dead matter

depression: a place that is lower than its surroundings

diverse: made of many different things

ecosystem: a community made of living things and the places in which they live

environments: the nonliving and living surroundings of organisms

erodes: wears away

herbivores: animals that eat only plants

hibernate: sleep or rest for a long period of time

omnivores: animals that eat both meat and plants

organisms: living things, such as plants and animals

predators: animals that hunt other animals

runoff: water from the land draining into a body of water such as a river or ocean

salt water: water holding large amounts of salt

settlements: places that are permanently occupied by people

Index

Get the best of both worlds.

AV2 bridges the gap between print and digital.

The expandable resources toolbar enables quick access to content including **videos**, **audio**, **activities**, **weblinks**, **slideshows**, **quizzes**, and **key words**.

Animated videos make static images come alive.

Resource icons on each page help readers to further **explore key concepts**.

Published by AV2
14 Penn Plaza, 9th floor
New York, NY 10122
Website: www.av2books.com

Library of Congress Control Number: 2020014387

ISBN 978-1-7911-2811-1 (hardcover)
ISBN 978-1-7911-2812-8 (softcover)
ISBN 978-1-7911-2813-5 (multi-user eBook)
ISBN 978-1-7911-2814-2 (single-user eBook)

Printed in Guangzhou, China
1 2 3 4 5 6 7 8 9 0 24 23 22 21 20

052020
101319

Project Coordinator: John Willis
Designer: Ana María Vidal

Every reasonable effort has been made to trace ownership and to obtain permission to reprint copyright material. The publishers would be pleased to have any errors or omissions brought to their attention so that they may be corrected in subsequent printings.

AV2 acknowledges Dreamstime, Getty Images, iStock, Minden Pictures, and Shutterstock as its primary image suppliers for this title.